YOUR KNOWLEDGE HAS VALUE

Rahul Ramachandran

Identification of low abundant protein biomarkers using blood as the starting sample

GRIN Verlag

Bibliografische Information der Deutschen Nationalbibliothek:

Die Deutsche Bibliothek verzeichnet diese Publikation in der Deutschen National-bibliografie; detaillierte bibliografische Daten sind im Internet über http://dnb.d-nb.de/ abrufbar.

Imprint:

Copyright © 2010 GRIN Verlag GmbH
Druck und Bindung: Books on Demand GmbH, Norderstedt Germany
ISBN: 978-3-656-63196-5

This book at GRIN:

http://www.grin.com/en/e-book/271078/identification-of-low-abundant-protein-biomarkers-using-blood-as-the-starting

GRIN - Your knowledge has value

Der GRIN Verlag publiziert seit 1998 wissenschaftliche Arbeiten von Studenten, Hochschullehrern und anderen Akademikern als eBook und gedrucktes Buch. Die Verlagswebsite www.grin.com ist die ideale Plattform zur Veröffentlichung von Hausarbeiten, Abschlussarbeiten, wissenschaftlichen Aufsätzen, Dissertationen und Fachbüchern.

Visit us on the internet:

http://www.grin.com/

http://www.facebook.com/grincom

http://www.twitter.com/grin_com

Identification of low abundant protein biomarkers using blood as the starting sample

Blood is the most popular biofluid used in proteomics to identify disease biomarkers. The potential of blood plasma or serum in diagnosis can be realized by the large amount of research and money spend on their study.

Interestingly, blood is a treasure trove of disease biomarkers. Every cell in the body leave a record of its physiological state in the form of waste or signal molecules in the products that it sheds to blood.. Hence, a small sample of blood could reveal the ongoing physiological and pathological states of tissues in the body(Liotta et al., 2003).

Important considerations for using blood as a starting sample

The first important consideration to be taken into account is the difference between plasma and serum. Plasma is the liquid component of the blood in which blood cells are suspended whereas, serum is protein solution left after the bulk of fibrinogen has been removed by conversion into fibrin clot, together with platelets. In addition, varying amount of other proteins also gets removed with fibrin clot by specific or non-specific interactions. And hence, the protein concentration of serum is less than plasma (Lum and Gambina, 1974, Landenson et al.: cited by Lundblad, 2005). Furthermore, the process/storage containers, time of clot retraction, centrifugation speed and temperature of storage also influence serum quality. The time between venipuncture and freezing, process/storage containers, centrifugation speed and temperature of storage are critical variables for plasma. Even though, it is difficult to control all these variables, but a standard operating procedure for blood collection could ensure reproducibility of the results(Lundblad, 2005).

Another factor to be considered is the dynamic qualitative and quantitative range of proteins in the blood. It is speculated that the dynamic range of protein concentration would be at least 10^6 or greater(Lundblad, 2005). Both serum and plasma have high protein content. But 99% of the blood protein consist of 22 proteins, major one being albumin, transferin, immunoglobulin and complement factors(Veenstra et al., 2005) (Figure 1). This dynamic range of proteins exceeds the analytical capability of current proteomic methods, thereby making detection of low abundance protein fraction which contains most of the undiscovered biomarkers extremely challenging(Tirumalai et al., 2003). The heterogeneity of plasma and serum makes prefractionation to remove highly abundant proteins, essential in proteomic analysis(Lundblad, 2005).

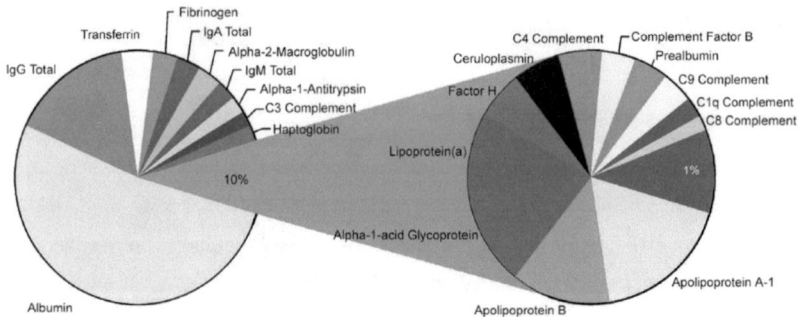

Figure 1: Pie-chart showing relative contribution of proteins within plasma(Tirumalai et al., 2003).

The various steps involved in detection of low abundance proteins in blood samples are summarized below:

Prefractionation

Prefractionation is done to reduce the dynamic range of protein concentration for detecting lower abundance proteins in blood samples. It is done prior to 2-DE and MS analysis. Some of the major prefractionation techniques are discussed below.

Electrophoresis based methods:

- Multicomponent electrolyzers (MCE) with isoelectric membranes: MCEs with isoelectric membranes is a preparative version of IPG. They are used to purify proteins in a liquid vein by capturing them in an isoelectric trap formed by two immobiline membranes having a pI range that include the pI of the protein species under study. When this apparatus was used for the analysis of human plasma with a pH range of 3-6, an increase in acidic proteins was seen when compared to the whole plasma(Herbert and Righetti, 2000).

- Gradiflow™: It's a preparative electrophoretic system that separates proteins on the basis of size and pI. The charge of the protein is controlled by the running buffer. Thus, when pH is below pI of the protein, a net positive charge develops. On the other hand, when the pH is above pI, protein gains a negative charge. The size separation is achieved by polyacrylamide membranes of specific pore sizes. Fitzgerald and Walsh (2010) used a prefractiontion device MF-10 that employed Gradiflow™ technology to separate plasma proteins. They were able to compartmentalize highly abundant proteins, thereby increasing the visualization of lower abundance proteins by 2-DE.

Affinity based methods:

Affinity based prefractionation could be used not only for reducing the complexity of protein sample, but also for enriching biomarkers. Some of these methods are briefly described.

- High abundant protein depletion: It is the most common prefractionation method used in proteomics (Righetti et al., 2005). Highly abundant albumin can be removed by a immobilized dye, Cibacorn Blue F3GA (Leatherbarrow and Dean, 1980) or specific immunosorbents (Tirumalai et al., 2003, Mehta et al., 2003). And IgG can be removed using Protein A(Lindmark et al., 1983).

- Cysteine containing peptide enrichment: In this technique thiol affinity resin is used to enrich cysteine containing peptides. But 95% of cysteine containing peptides are contributed by two major abundant proteins: albumin and transferring (Pernemalm et al., 2009). By depleting high abundant proteins, Liu et al. (2006, cited by Pernemalm et al., 2009) identified high number of proteins (n =1977) using cysteine enrichment.

Chromatographic methods:

Solid-phase chromatographic techniques are used in proteomics. By selecting appropriate adsorbents and conditions for adsorbtion-elution, low or high selectivity for proteins implemented. The chromatographic techniques can be used as a single unit or in combinations which are termed two or multidimensional chromatography. A variety of chromatographic techniques have been used to reduce the complexity of samples: anion exchange (Wiesner, 2004, Zhang et al., 2004), cation exchange (Tirumalai et al., 2003), size exclusion chromatography and lectin affinity chromatography (Geng et al., 2001). Recently, Mahn et al. (2010) used hydrophobic interaction chromatography (HIC) to analyze blood plasma. They were successful in reducing the complexity, however, as most of highly abundant proteins have medium hydrophobicity, complete depletion didn't occur. Thus, further optimization is necessary for improving its efficiency.

Protein separation after prefractionation

Two-dimensional Polyacrylamide Gel Electrophoresis (2-DE)

This technique resolves protein in first dimension by isoelectric point and in second dimension by molecular weight. It uses immobilized pH gradient (IPG) strips consisting of polycarboxylic acid ampholytes that reproducibly create stable pH gradients. The 2-DE system utilizes IPG strips with immobilized polycarboxylic ampholytes and appropriate hardware to transfer proteins from IPG strips to SDS-PAGE gels.

2-DE can separate thousands of proteins and allows comparative protein profiling between different biological samples(Liebler, 2002). Nevertheless, 2-DE is incapable of detecting low abundant, high molecular weight, basic and hydrophobic proteins. These challenges have been overcome by recent progress in 2-DE technologies. Inagaki and Katsuta (2004) have reviewed some of the methods to improve protein separation by 2-DE. These include, sample prefractionation techniques, some of which have been discussed earlier. Others like use of commercially available nrIPGs, longer IEF gels, multiple and longer SDS-PAGE gels improved protein separation in their respective dimensions.

Two-dimensional Liquid Chromatography (2-DLC)

2DLC profiling of proteins can be done by ion-exchange chromatography(Lubman et al., 2002) or chromatofocussing (Whitelegge, 2005) to separate proteins based on their charge, followed by Reverse-Phase Liquid Chromatogarphy-Electrospray Mass Spectrometry to separate proteins based on hydrophobicities in second dimension and measure their exact M_r (Sahab et al., 2007).

Sample clean up prior to MS

Sample clean up is essential before MS analysis because some of the charged and ionisable molecules in the sample can interfere with ionization of analyte causing signal suppression and / or increase in background noise. Some of the sample clean up techniques available is summarized in the table below:

Molecules to eliminate	Technique	Further clean-up
SDS	HILIC/SCX	Reverse-phase
Non-ionic detergents	SCX	Reverse-phase
Zwitterionic detergents	SCX or Reverse phase C_{18}, C_8, C_4	Reverse-phase
Salts	Reverse phase C_{18}, C_8, C_4	-
Chaotropes (urea)	Reverse phase C_{18}, C_8, C_4	-
Buffers	Reverse phase C_{18}, C_8, C_4	-

HILIC: Hydrophobic interaction SCX: Strong cationic exchange chromatography

(Source: Quadroni, 2010)

Mass spectrometry

MALDI-TOF-MS:

For MALDI-TOF-MS, the sample is mixed with a matrix and spotted on a plate, allowing it to dry. A short pulse of light from a laser is shined on the plate, following which, the molecules in the matrix absorbs the energy from the light; this energy is utilized to volatize the sample molecules. The sample ions are accelerated though a flight tube using high voltage and they separate according to the kinetic energy. After the ions reaches the detector, next pulse of ions enters the flight tube and thus mass spectrum is generated from sum of spectra recorded for several pulses of ions(Counterman et al., 2003).

MALDI-TOF-MS was used by Finnskog et al. (2006) in combination with porous silicon nanovial arrays to identify serum biomarkers. Studies by Ahmed et al. (2005) and Avasarala et al. (2005) identified biomarkers for ovarian cancer and multiple sclerosis respectively, using MALDI-TOF-MS.

Advantages: MALDI-TOF peptide fingerprinting takes less time. And the subsequent database search takes 3-4 minutes. Thus, it's rapid and straightforward. The operation of MALDI-TOF-MS is easy compared to ESI-MS/MS.

Electrospray ionization tandem mass spectrometry (ESI-MS/MS):

In ESI-MS/MS analysis, the sample is sprayed using electrospray needle held at high potential and the charged droplets thus produced will pass through a heated capillary or a countercurrent gas flow, and get desolvated. The ions formed are then analysed by the mass spectrometer. The MS/MS spectrum obtained is searched across databases and the protein is identified by multiple independently sequenced from the same protein.

Tirumalai et al. (2003) identified 340 serum proteins using ESI-MS/MS coupled with reverse-phase chromatography. The major advantages of using ESI-MS/MS are: high sensitivity and reliability.

Database search

- *Mascot (http://www.matrixscience.com):* It uses probability based scoring. In this scoring method, the probability of the observed match between the experimental data and calculated values to occur as a random/chance event is calculated. Hence, the lowest probability gives the best match.

- *Profound* (http://bioinformatics.genomicsolutions.com/service/prowl/profound.html): Profound also based probability-based scoring. It uses bayesian algorithm to identify proteins based on several criteria like peptide cleavage pattern, observed mass values etc.

References

AHMED, N., OLIVA, K. T., BARKER, G., HOFFMANN, P., REEVE, S., SMITH, I. A., QUINN, M. A. & RICE, G. E. 2005. Proteomic tracking of serum protein isoforms as screening biomarkers of ovarian cancer. *Proteomics*, 5, 4625-36.

AVASARALA, J., WALL, M. & WOLFE, G. 2005. A distinctive molecular signature of multiple sclerosis derived from MALDI-TOF/MS and serum proteomic pattern analysis. *Journal of Molecular Neuroscience*, 25, 119-125.

COUNTERMAN, A. E., THOMPSON, M. S. & CLEMMER, D. E. 2003. Identifying a Protein by MALDI-TOF Mass Spectrometry: An Experiment for the Undergraduate Laboratory. *Journal of Chemical Education*, 80, 177

FINNSKOG, D., JARAS, K., RESSINE, A., MALM, J., MARKO-VARGA, G., LILJA, H. & LAURELL, T. 2006. High-speed biomarker identification utilizing porous silicon nanovial arrays and MALDI-TOF mass spectrometry. *Electrophoresis*, 27, 1093-103.

FITZGERALD, A. & WALSH, B. J. 2010. New method for prefractionation of plasma for proteomic analysis. *ELECTROPHORESIS*, 31, 3580-3585.

GENG, M., ZHANG, X., BINA, M. & REGNIER, F. 2001. Proteomics of glycoproteins based on affinity selection of glycopeptides from tryptic digests. *J Chromatogr B Biomed Sci Appl*, 752, 293-306

HERBERT, B. & RIGHETTI, P. G. 2000. A turning point in proteome analysis: sample prefractionation via multicompartment electrolyzers with isoelectric membranes. *Electrophoresis*, 21, 3639-48.

INAGAKI, N. AND KATSUTA, K. 2004. *Large gel two-dimensional electrophoresis: Improving recovery of cellular proteome*, Curr. Proteomics, 1(1) 35-39.

LEATHERBARROW, R. J. & DEAN, P. D. 1980. Studies on the mechanism of binding of serum albumins to immobilized cibacron blue F3G A. *Biochem. J.*, 189, 27-34.

LIEBLER, D. 2002. *Introduction to proteomics: tools for the new biology*, Humana Press.

LINDMARK, R., THORÉN-TOLLING, K. & SJÖQUIST, J. 1983. Binding of immunoglobulins to protein A and immunoglobulin levels in mammalian sera. *J Immunol Methods*, 62, 1-13.

LIOTTA, L. A., FERRARI, M. & PETRICOIN, E. 2003. Clinical proteomics: Written in blood. *Nature*, 425, 905-905.

LUBMAN, D. M., KACHMAN, M. T., WANG, H., GONG, S., YAN, F., HAMLER, R. L., O'NEIL, K. A., ZHU, K., BUCHANAN, N. S. & BARDER, T. J. 2002. Two-dimensional liquid separations-mass mapping of proteins from human cancer cell lysates. *J Chromatogr B Analyt Technol Biomed Life Sci*, 782, 183-96.

LUNDBLAD, R. L. 2005. Considerations for the Use of Blood Plasma and Serum for Proteomic Analysis. *The Internet Journal of Genomics and Proteomics* [Online], 1. Available: http://www.ispub.com/ostia/index.php?xmlFilePath=journals/ijgp/vol1n2/plasma.xml.

MAHN, A., REYES, A., ZAMORANO, M., CIFUENTES, W. & ISMAIL, M. 2010. Depletion of highly abundant proteins in blood plasma by hydrophobic interaction chromatography for proteomic analysis. *Journal of Chromatography B-Analytical Technologies in the Biomedical and Life Sciences*, 878, 1038-1044

MEHTA, A. I., ROSS, S., LOWENTHAL, M. S., FUSARO, V., FISHMAN, D. A., PETRICOIN, E. F. & LIOTTA, L. A. 2003. Biomarker amplification by serum carrier protein binding. *Dis Markers*, 19, 1-10.

PERNEMALM, M., LEWENSOHN, R. & LEHTIÖ, J. 2009. Affinity prefractionation for MS-based plasma proteomics. *PROTEOMICS*, 9, 1420-1427.

QUADRONI, M. 2010. *Protein separation and sample preparation for mass spectrometry* [Online]. Available: http://www.ch.embnet.org/CoursEMBnet/PROT04/Slides/Prot_sep%26sample_prepMS.pdf [Accessed 19 Dec 2010].

RIGHETTI, P. G., CASTAGNA, A., ANTONIOLI, P. & BOSCHETTI, E. 2005. Prefractionation techniques in proteome analysis: The mining tools of the third millennium. *ELECTROPHORESIS*, 26, 297-319.

SAHAB, Z. J., SEMAAN, S. M. & SANG, Q. X. 2007. Methodology and applications of disease biomarker identification in human serum. *Biomark Insights,* 2, 21-43.

TIRUMALAI, R. S., CHAN, K. C., PRIETO, D. A., ISSAQ, H. J., CONRADS, T. P. & VEENSTRA, T. D. 2003. Characterization of the low molecular weight human serum proteome. *Mol Cell Proteomics,* 2, 1096-103.

VEENSTRA, T. D., CONRADS, T. P., HOOD, B. L., AVELLINO, A. M., ELLENBOGEN, R. G. & MORRISON, R. S. 2005. Biomarkers: mining the biofluid proteome. *Mol Cell Proteomics,* 4, 409-18

WHITELEGGE, J. 2005. Tandem mass spectrometry of integral membrane proteins for top-down proteomics. *TrAC Trends in Analytical Chemistry,* 24, 576-582.

WIESNER, A. 2004. Detection of tumor markers with ProteinChip technology. *Curr Pharm Biotechnol,* 5, 45-67.

ZHANG, Z., BAST, R. C., YU, Y., LI, J., SOKOLL, L. J., RAI, A. J., ROSENZWEIG, J. M., CAMERON, B., WANG, Y. Y., MENG, X. Y., BERCHUCK, A., VAN HAAFTEN-DAY, C., HACKER, N. F., DE BRUIJN, H. W., VAN DER ZEE, A. G., JACOBS, I. J., FUNG, E. T. & CHAN, D. W. 2004. Three biomarkers identified from serum proteomic analysis for the detection of early stage ovarian cancer. *Cancer Res,* 64, 5882-90.